Yi Min's Great Wall

A Materials Engineering Story

Written by the Engineering is Elementary Team
Illustrated by Jeannette Martin

Chapter One | The Bandit Bunny

Yi Min peered through tall bamboo reeds into the schoolyard garden. It was hard to sit like a statue, not moving a muscle, but still she didn't budge. Her mission was too important. Yi Min was acting as an undercover detective—an undercover bunny detective.

Ever since spring had arrived in Yi Min's village, just outside of Beijing, she and her classmates had gone to their garden once a week to pull weeds and record the growth of their plants. But as the weeks went on, there were fewer and fewer plants. A rabbit, or *tuzi*, kept eating the vegetables they had planted. Yi Min had decided to save the garden by catching this bandit *tuzi*.

She imagined the thief: a soft brown rabbit wearing a small black mask over his eyes. He would hop slowly and quietly into the garden and munch away. Yi Min would jump out from behind the bush and yell, "*Ni hao*!—hello!" The bunny would be too scared to come back. Yi Min's whole class would thank her for saving their garden.

Yi Min yawned. *The problem with being an undercover bunny detective*, she thought, *is that sitting still like a statue is no fun!* Maybe her friend Chen could come help her. Yi Min stood and stretched her sore muscles. Her detective work would have to wait until tomorrow.

Chapter Two | A Special Trip

Yi Min woke up excited the next day. Maybe this would be the day she finally scared away the bunny! She jumped out of bed and looked at her calendar—and then got excited for another reason. Today she was going on a special trip with her class. They were going to visit the Great Wall of China.

The bus trip was long, and Yi Min was tired. But as she and her classmates stood in front of a small section of the wall, Yi Min felt amazed. She looked into the distance and saw the wall stretching like a thick vine across the countryside for miles and miles.

"Class, who can tell me something about the emperor who built the Great Wall?" *Lao Shi*, their teacher, asked.

Yi Min raised her hand. "His name was Ying Zheng," she answered. "Ying Zheng was very young when he became emperor, but he was fierce and strong. He ordered many people to leave their homes and farms to work on the wall. If I were emperor, I wouldn't have been so impatient. And I would have made the wall rainbow colors. And . . ."

Yi Min's classmates began to giggle. Yi Min felt her face get hot and could tell it was turning red. She knew that sometimes her imagination carried her away, but she couldn't help it when all of the ideas she had began to bubble out of her head.

Lao Shi called the class to order and pointed towards the wall behind her. “This wall was built to protect the people of our country. Who can tell me some other reasons that people build walls?”

“To keep out the cold and the rain,” answered Jing.

“To keep warm air inside,” said Tao.

“To keep animals outside,” said Chen.

Yi Min giggled. She imagined a bear outside the wall of her bedroom tapping on the window, hungry for dried fruit

and mooncakes. She would invite him in and ask him if he wanted to stay for tea. *Do bears like honey in their tea?* she wondered.

"Yi Min," *Lao Shi* called, interrupting her daydream. "Do you know why the Great Wall of China is called the Sleeping Dragon?"

Yi Min thought for a moment and looked at the Great Wall in front of her. "The wall is long and large like a dragon, and it protected the people of China from enemies and wars, just like a fierce dragon would."

"Very good," *Lao Shi* said. "I want all of you to draw some pictures and write down what you see here today."

Yi Min moved to get a closer look at the wall. It was so tall that when she stood next to it she could hardly see the top. She could almost feel the great emperor Ying Zheng there with her as she drew a picture of the wall.

Chapter Three | A New Plan

The next day at school, it was time for another visit to the garden. As Yi Min walked towards the plants, Chen wandered over to join her.

"Come on, let's check on the cabbage," he suggested.

Yi Min and Chen found themselves looking at a ragged, half-eaten cabbage plant.

"Oh, no!" Chen cried. Other students came running.

"That *tuzi* is ruining our garden!" Ping exclaimed.

"I wish there was a way to get rid of it," Ha Young added.

Lao Shi walked over to the group. "I know you're sad that your plants are getting eaten, but the *tuzi* doesn't mean to upset you. He is just looking for food. I brought some pea seedlings with me. Why don't you plant these in that clear patch of garden? Maybe these plants won't be as tempting for the rabbit."

The students headed toward a clear patch of soil in the back of the garden. Yi Min and Chen lingered behind.

"We need to think of a way to stop that rabbit," Chen said.

"I tried to!" Yi Min said. "I came to the garden to be a bunny detective and scare him away. Only the bandit bunny never came."

"A bunny detective?" Chen asked skeptically. "There has to be a better way to protect the plants."

Yi Min thought for a moment. "You're right," she said. "Instead of scaring away the bunny, we should think about protecting the plants. And I know just how we can do it: we can build our own Great Wall of China!"

"Yi Min, you can't be serious," Chen said. "Do you know how hard that would be?"

Yi Min smiled. Her mind was spinning as she pictured different possibilities. "Don't be silly," she said. "It doesn't have to be as big or as long as the real Great Wall. We'll work on it after dinner. But don't tell anyone!" She could just imagine the glory of announcing to her classmates her victory over the bunny. "It will be great!" Yi Min skipped off to help plant the seedlings.

Chapter Four

A Talk with Grandfather

Yi Min walked home from school that day thinking about the wall she and Chen could build. She knew one person who had helped to design real structures like bridges and buildings: her grandfather. He was an engineer—someone who used his knowledge of science and math, along with his creativity, to design things that solved problems. *I'll have to ask Grandfather more about the things he helped design*, Yi Min thought.

At home, Yi Min found her grandfather sitting outside on the patio sipping green tea out of a jar.

"Hello, Grandfather," she said. "I have a question for you. Yesterday in school we visited the Great Wall of China. Did you build structures like that?"

"I helped build things, but not with a hammer and nails," Grandfather said. "I was a materials engineer. I helped design new technologies—new materials with new properties that were then used to build things. Do you remember when I showed you the bridge over the Yangtze River that I worked on?"

"But that bridge didn't look like it was made of any fancy new technology," Yi Min said. She imagined an electronic bridge with a road that moved cars along and a clear bridge made of futuristic crystals.

Grandfather interrupted Yi Min's thoughts. "Technology is what people create and use to solve a problem. It could be a way of doing things, or an object, or even a material. I helped design the mortar for the bridge. It's an important technology because it's the part that sticks the bricks together. Before I began to work on the bridge, the mortar had started to crumble. I helped to create a new mortar that was strong and would last a long time."

"You must know about lots of different materials," said Yi Min.

"I do," Grandfather said. "If you take some time to look at the walls and structures around you, you can learn about materials as well. Was that your question—what kind of work I did?"

"Not really," Yi Min admitted. "I need your help. We have a garden at school and we want to take good care of it so we can harvest our vegetables. But there's a *tuzi* that keeps coming in and eating our plants! I thought I might build a wall to keep the bunny out—but I don't know where to start."

"Maybe you should use the engineering design process," Grandfather said. "That's how most engineers solve problems. You've already started the process by asking questions. You'll need to imagine lots of solutions, make a plan, create it, and then improve it to make it better."

"Oh, yeah," Yi Min said. "I remember you telling me about the engineering design process when you showed me the Yangtze bridge."

Grandfather patted Yi Min on the shoulder and picked up his cane. "With your imagination and creativity, I know you'll be able to build a wonderful wall," he said.

Yi Min began to think about all of the different types of walls around her. The walls in her house were made out of wood. The walls of her school were made out of brick. Walls could also be made of stone, glass, or metal. The walls of the blanket forts she built were made out of cloth.

The key, thought Yi Min, *must be to match the material to the job you need it to do. A cloth wall wouldn't work very well out in the garden because it would get soggy in the rain. And I bet the bandit bunny would nibble his way through,* she thought. *But bricks and mortar, the materials Grandfather used for the Yangtze bridge . . . maybe that could work.*

1
2
3
2
2
1
FIX the BRIDGE
2
THE ENGINEERING DESIGN PROCESS

Chapter Five

Beginning to Engineer

After dinner, Chen and Yi Min met in the school garden.

"Well," Chen said, "what do we do now?"

Yi Min imagined that she was the emperor's lead engineer, ready to create a plan for building the Great Wall. She thought back to movies she had seen and tried to be like a general inspiring her troops.

"Chen," she began, "our mission is to create a wall that stops the bandit bunny from eating our plants. It may be hard work. It may be dangerous work."

"Dangerous work?" asked Chen.

"Okay, maybe not dangerous," said Yi Min, "but it might take us some time to figure out how to build the wall. I was thinking about all the materials that walls are made of—like wood and stone and bricks."

"How would we get bricks?" Chen asked. "It could be expensive."

"Buying bricks could be expensive," Yi Min agreed. "But gathering stones near the river wouldn't."

"That's true," Chen said, beginning to smile. "But how would we hold the stones together?"

"That's where the mortar comes in," Yi Min said. "Mortar is a mixture that can be used to stick bricks or stones together. We have plenty of soil in the garden and water in the stream. We could try using that to make mortar."

Chen and Yi Min scooped some soil into a bucket. Chen looked at the soil Yi Min had collected.

"There are some weeds and sticks in here," he pointed out. "We better get rid of them."

Yi Min reached to pull the sticks out, but stopped. "Wait," she said. "How do we know we should get rid of them? What if they make our mortar stronger?"

"I don't see how those sticks will help make a better wall," Chen protested.

"I'm not saying you're wrong. But we're not sure," Yi Min said. "Let's make a bunch of different kinds of mortar: mud with sticks, plain mud, sand, and maybe even some from the clay from the stream banks. Then we can use the mortar to make little walls to test. We can compare the walls after they're dry."

Yi Min added water to the dirt and sticks in her bucket and she and Chen began to make a wall from the mud mortar and stones. After they were through with that batch, Chen grabbed a bucket and walked towards a sandy patch of garden. He scooped up some of the sandy dirt, then added water. Yi Min and Chen used that mixture and more stones to make another wall. Soon they had rows of miniature walls baking in the sun.

"I think my grandfather would be proud of us," Yi Min said. "We experimented with materials; next we can test the mortars to see which ones make the strongest walls."

Chen smiled. *This must have been what it was like to work for one of the great emperors*, Yi Min thought. She couldn't wait to begin building their wall.

Chapter Six | Testing and Improving

The next day after school Yi Min and Chen returned to the garden.

"I brought a notebook in case we need to write anything down," said Yi Min. "And some green tea, to keep up our strength," she added. She plopped down and reached out to poke one of the walls. It was dry and hard.

"This is going to be great," she said. She reached to pat another wall, but as soon as she touched it, the stones tumbled to the ground.

"Don't knock down the walls, Yi Min!" warned Chen.

"I didn't mean to," Yi Min said. "It just fell apart."

Chen looked down at the crumbled wall. "That's one of the walls we made with sand mortar."

"Hmm. The materials are different," said Yi Min. "This reminds me of something my grandfather tells me. He tells me to listen to the earth."

Chen bent down and put his ear to the ground. "Like this?" he asked.

"No, silly. Hold on, I'll show you."

Yi Min ran towards the stream at the edge of the garden patch and began to scoop rocks into a bowl. Once the bowl was full, she brought her treasures back to Chen.

"Each rock has its own personality. This rock is hard and strong, just like a brave soldier. This rock is frail and brittle, like a wise elder. All of these rocks have different properties that make them good for some things and bad for other things—just like the mortars we've made. Some materials make mortars that are good for building a wall and others do not!"

"The sand mortar that fell apart wasn't very strong," Chen said.

"Right," agreed Yi Min. "But I wonder if we can mix the sand with something else so that we can use it—like this clay soil near the stream. It's gooey, like glue," Yi Min said.

Yi Min and Chen began imagining lots of different mixtures of sand, soil, and clay, with different amounts of each material in each mixture. Yi Min began to write down the mixtures they had already made that worked well, and all of the new

mixtures they planned to create. Yi Min and Chen mixed many different mortars.

Yi Min was sure that her grandfather would be proud of the materials engineering that she was doing.

Chapter Seven | Revealing the Great Wall

Yi Min and her classmates continued to study the Great Wall of China in school the next week. After school, Yi Min and Chen tested their mortars and decided which mixture was the best choice. Each day they gathered more stones, mixed more and more mortar, and made their wall around the garden bigger and bigger. The day before the class was going to make their next visit to the garden, Yi Min and Chen finished building the last section of the wall with sticky mortar. Their very own great wall was complete.

The next day in class, Yi Min couldn't wait for her classmates to see what she and Chen had created. Finally, *Lao Shi* said the words Yi Min had been waiting for.

"Why don't we go outside and tend the garden?"

None of Yi Min's classmates seemed excited to check on the seedlings they had planted.

Tao raised her hand. "Nothing that we plant gets to grow. The rabbit eats it too quickly."

Chen turned and looked at Yi Min. She smiled back at him. She was so excited, she could barely sit still.

Lao Shi nodded. "You are right. We will have to work on a way to keep our plants safe. For now, though, we will continue to water and weed the garden. Let's go."

Yi Min grabbed Chen's hand and pulled him to the front of the line. She wanted to make sure she didn't miss a moment.

Outside, Yi Min stood by the bamboo reeds on the side of the garden. She wanted to see everyone's reaction.

The class spread out in the garden. Some students pointed towards half-eaten plants, and others lifted leaves with nibble marks on them. Some of the students made their way towards the back of the garden to check on the newly planted seeds.

"Look!" one of the girls cried. "There's a wall around the new seedlings! They're all healthy and growing!"

More students headed towards the back of the garden, gathering around the plants. *Lao Shi* followed them over. Yi Min stood back proudly, looking from her classmates to Chen.

"This is wonderful," *Lao Shi* said. "Who built this wall around our new plants?"

The children in the class looked at one another. Then a loud voice coming from behind them caught their attention.

"We did!"

Lao Shi and the class turned around to see Yi Min and Chen standing behind them, grinning at the great wall they had built.

"Yi Min, Chen, you two have been wonderful engineers. Maybe you can teach your classmates what you have learned. We could all work together to make a wall around the other plants in the garden," *Lao Shi* said.

Chen began telling the class about all the steps of the engineering design process that the two of them had used. Yi Min explained how they had discovered that some types of mortar were stronger than others. As Yi Min spoke with her classmates, she realized that one very special person was missing from this discussion about materials and engineering.

Chapter Eight | A Real Engineer

Later that day, Yi Min held Grandfather's hand as they walked into the school yard. She pointed towards the wall that she and Chen had created.

Grandfather placed his hand on Yi Min's shoulder as he surveyed her work. "It's just like a miniature sleeping dragon, Yi Min," he said. "You have done a very good job."

"Building the wall and designing the mortar was such hard work, Grandfather," Yi Min said. "It made me think about how much soil and how many rocks were used to build the Great Wall. I can hardly imagine how they put everything together and made the wall

go on for so many miles."

"I can imagine it," said Grandfather, smiling down at her. "They just needed a good engineer to lead them. Just like you, Yi Min!"

Yi Min smiled back. She did feel like an engineer.

Design a Mortar Mixture

Have you ever made a wall from blocks? Different kinds of materials make very different kinds of walls. Using what you learned from Yi Min, your goal is to design a mortar mixture from earth materials that will hold a stone wall together.

Materials

- ☐ Small stones
- ☐ Twigs, toothpicks, or craft sticks
- ☐ Earth clay (really sticky dirt)
- ☐ Potting soil
- ☐ Straw or grass
- ☐ Sand
- ☐ Bowl
- ☐ Mixing spoon

Ask About Your Mortar Design

What earth materials make a strong and sticky mortar mixture? Build three small walls using mortars made of wet sand, wet soil, and wet clay. Let the walls dry for a few days. Do any of the earth materials make good mortars on their own? Can you imagine ways to combine the materials so that they make a better mortar? What might happen if you built a wall with a mortar mixture of straw, clay, and a few drops of water? What about a mortar mixture made from potting soil, sand, and clay?

Build Your Wall

Out of all the mixtures you imagined, choose one to plan and create. Make a list of the materials you'll need to create your mortar. Then think about how to construct your wall. Make a drawing of how you want your wall to look. Do you think you could use twigs to make your wall stronger and taller? Where will you put the mortar? Once you have your plan ready, create your wall!

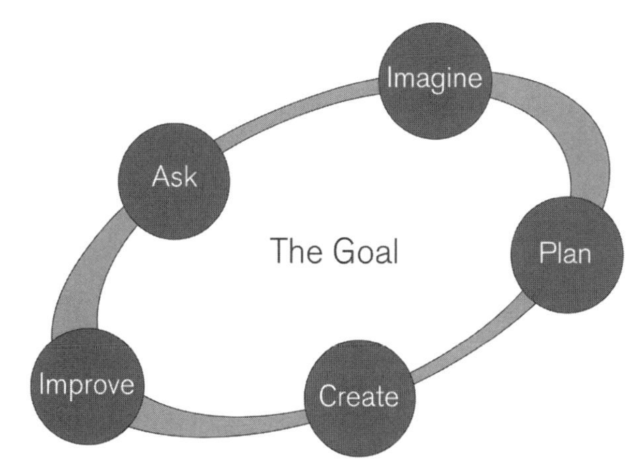

Test Your Mortar Design

Let your wall dry before you test how strong it is. Once it's dry, try rolling a heavy ball, like a baseball, against it. Is your mortar strong enough to stay upright, or does it fall over? Does your wall stick together, or does it crumble? If it breaks, where or how does it break?

Improve Your Mortar

Use the engineering design process to improve your mortar and wall. Use what you learned from how your wall broke to redesign your mortar mixture. Can you design a mortar that makes your wall taller or stronger? Take a walk around your neighborhood. What types of walls do you see? How are they made? Could any of the designs you see inspire the design of your wall?

Glossary

Clay: An earth material made of tiny particles of rock that pack tightly together. Clay is sticky when wet and hardens when it dries.

Design: A plan for a solution to a problem.

Engineer: A person who uses his or her creativity and understanding of mathematics and science to design things that solve problems.

Engineering design process: The steps that engineers use to design something to solve a problem.

Experiment: To conduct a test.

***Lao shi*:** Mandarin Chinese word for teacher. Pronounced *LAO-sure*.

Material: Something you can use to create things.

Materials engineer: A person who uses his or her creativity and knowledge of science and math to solve problems related to materials and create new materials with new properties.

Mixture: Two or more materials that are combined together.

Mooncake: A Chinese pastry made from a thick, sweet filling surrounded by a thin crust.

Mortar: A mixture that is used to make stones, rocks, or bricks stick together.

Ni Hao**:** Mandarin Chinese word for hello. Pronounced *NEE-how*.

Properties: Things about a material, like how it looks and feels or how it behaves when wet, that make it different from other materials.

Solution: The answer to a problem, or a design that fixes the problem.

Technology: Any thing, system, or process that people create and use to solve a problem.

Tuzi**:** Mandarin Chinese word for rabbit. Pronounced *TOO-zih*.

Yangtze: A river than runs from Tibet through China and into the East China Sea. It is the longest river in China. Pronounced *YANG-see*.